SCIENTIFIC AMERICAN EDUCATIONAL PUBLISHING

THINK OUTSIDE THE BOX

10 FUN PROJECTS USING ENGINEERING

BRING SCIENCE HOME

Published in 2025 by The Rosen Publishing Group, Inc.
2544 Clinton Street, Buffalo, NY 14224

Contains material from Scientific American, a division of Springer Nature America, Inc., reprinted by permission as well as original material from The Rosen Publishing Group.

First Edition

Editor: Kristen Rajczak Nelson
Designer: Rachel Rising

Activity on p. 5 by Science Buddies, Sabine De Brabandere (January 23, 2020); p. 11 by Science Buddies, Ben Finio (October 5, 2017); p. 17 by Science Buddies, Ben Finio (February 4, 2016); p. 21 by Science Buddies (May 7, 2015); p. 25 by Science Buddies (December 6, 2012); p. 31 by Science Buddies, Sabine De Brabandere (October 31, 2019); p. 37 by Science Buddies, Svenja Lohner (December 19, 2019); p. 43 by Science Buddies (August 20, 2015); p. 49 by Science Buddies, Ben Finio (February 23, 2017); p. 55 by Science Buddies, Ben Finio (June 20, 2019).

All illustrations by Continuum Content Solutions

Photo Credits: pp. 3, 4, 17, 20, 25, 28, 37, 41, 55, 58 cve iv/Shutterstock.com; pp. 4, 5, 11, 17, 21, 25, 31, 37, 43, 49, 55 Anna Frajtova/Shutterstock.com.

Cataloging-in-Publication Data

Names: Scientific American, Inc.
Title: Think outside the box: 10 fun projects using engineering / edited by the Scientific American Editors.
Description: New York : Scientific American Educational Publishing, an imprint of Rosen Publishing, 2025. | Series: Bring science home | Includes glossary and index.
Identifiers: ISBN 9781725350694 (pbk.) | ISBN 9781725350700 (library bound) | ISBN 9781725350717 (ebook)
Subjects: LCSH: Engineering--Experiments--Juvenile literature. | Engineering--Juvenile literature.
Classification: LCC TA149.T45 2025 | DDC 620.0078--dc23

Manufactured in the United States of America

Some of the images in this book illustrate individuals who are models. The depictions do not imply actual situations or events.

CPSIA Compliance Information: Batch #CSSA25. For further information contact Rosen Publishing, New York, New York at 1-800-237-9932.

CONTENTS

THESE ACTIVITIES INCLUDE SCIENCE FAIR PROJECT IDEAS.

INTRODUCTION

Science and engineering go hand in hand—with lots of fun ways to see how engineering plays a role in the world around us! In this book, you'll see art and engineering come together as you make a mobile, create a machine that can copy your drawings, find the science behind sharing candy, and so much more!

Projects marked with ⚛ include a section called Science Fair Project Ideas. These ideas can help you develop your own original science fair project. Science fair judges tend to reward creative thought and imagination, and it helps if you are really interested in your project. You will also need to follow the scientific method. See page 61 for more information about that.

Lift a Large Load Using Liquids

LIQUID POWER: BUILD A LIFT THAT USES WATER TO HELP!

Not even the strongest human could lift a truck into the air. Our brains, however, are smart enough to create a tool that can lift heavy objects for us: hydraulic lifts! You can find them at car repair shops, in wheelchair lifts, and even on skyscraper construction sites. They use water (or other liquids) to increase the force available to lift things. And you will get to build one in this activity! Try it out and discover how much easier it is to lift with the power of water.

PROJECT TIME

30 to 45 minutes

KEY CONCEPTS

Physics
Engineering
Simple Machines
Force
Pressure

BACKGROUND

If you have ever used a wheelbarrow, you know it can help to lift heavy objects: lifting the handles of the wheelbarrow is easier than lifting the object itself. A hydraulic lift accomplishes the same task. It allows you to move a heavy object with much less effort.

"Hydraulic" refers to tools that operate by moving liquids, such as water or oil. You cannot compress liquids (unless you have special equipment). This means that no matter how hard you press on a liquid you cannot noticeably change its volume. Try it out! Fill an oral medicine syringe with water, close its open tip with a finger, and try to press the plunger in. You will see that you cannot. Even if you could place an elephant on the plunger, it would not compress the fluid in the syringe. So what will you be able to do with this waterpower?

In the effort to move something, it is not only important to exert force but also to consider the full area over which the force is spread. Think of a time when you played with Play-Doh. Pushing a finger into the dough is easy, but pushing the palm of your hand into the dough is harder—you probably felt more resistance. Because your hand has a larger surface area, your effort is spread out over a larger area, and the dough does not receive as much force for each square inch as it does with a single-finger push. Now think about this problem in reverse: sometimes you want less force per square inch, such as when you are trying to lift something. Is there a way to place an object on a larger surface so it creates less pressure when you're trying to lift it? Try the activity to find out!

MATERIALS

- Two stainless steel wall plates, at least 4.5 by 2.75 inches (11.4 by 7 cm) in size. (Available in hardware stores.)
- Two small oral medicine syringes (these are usually given in metric measurements—25 milliliters or 25 CC works well.) (Available at pharmacies.)
- Large oral medicine or irrigation syringe (one that is approximately 60 milliliters or 60 CC works well; one with a catheter tip attached will be easiest to work with). (Available at pharmacies.)
- Extra-strong wall-mount double-sided tape

- 4 feet (1.2 m) of silicone fuel tubing with a 5/32-inch (4 mm) inner diameter. (This is available in hobby shops, hardware stores, and online.)
- Scissors
- Hardware vise
- Sturdy surface to which the vise can be attached
- Jar full of water (with lid closed tightly) and/or other heavy object that will fit on the wall plate
- Large bowl of water
- Ruler
- Paper and pen or pencil
- Adult helper
- Workspace that can get messy
- Plumber's putty (optional) (available at hardware stores)
- Scale (optional)
- Other sizes of syringes with catheter tip (optional)

PREPARATION

- Put one small syringe away to the side. This syringe will be the "leading" syringe; you will use it to make the lift move. The other two syringes (one small one and the large one) will hold the weight that needs to be lifted. These will be the "helping" syringes.
- Use the extra-strong double-sided tape to adhere the center of one wall plate to the flat top of each of the plungers on the two helping syringes.
- Carefully cut the tubing in two pieces, each about 2 feet (0.6 m) long.
- For each helping syringe (the syringes with plates attached), connect a plastic tube to its tip.
- Push the plunger in each helping syringe all the way down so that it is fully depressed into the cylinder.
- One at a time, put the free end of each tube into the water, and pull the plungers smoothly up as far as they will go without falling out of the syringes. The syringes will suck up water.

- Hold each syringe (one at a time) with its tip pointing up. Remove the air bubble near the tip of the syringe by pushing the plunger in slowly until all the air is pushed out of the tube.
- Place the end of the small syringe's tube back in the water, and suck in some more water.
- Very little air should be left in either of the tubes or helping syringes. The small syringe should be full, and the larger one can be half-full.
- Attach the vise to a sturdy surface.

PROCEDURE

- Insert the tip of the leading syringe into the free end of the tube connected to the small helping syringe. Push the plunger of this helping syringe in. *What happens to the leading syringe? Why would this happen? Do you need to push hard to get the water to transfer from one syringe to the other?* (If the tubing pops off, refill the tubing and helping syringe with water, and reattach the tubing to the leading syringe. If needed, seal the connection with plumber's putty.)
- The leading syringe should now be full of liquid.
- Use the vise to carefully hold the helping syringe vertically with the plate horizontal and facing up. Be careful not to squish the cylinder of the syringe as this will obstruct the movement of the plunger.
- Fill the jar with water and place it in the precise center of the platform attached to the helping syringe in the vise. *Do you think you can easily lift this weight by pushing on the leading syringe?*
- Try it out. Push gently so the jar does not tumble off. You might need a helper to gently keep the construction upright. *Was your prediction correct?*

- Take the jar off the platform, push the plunger of the helping syringe down until the leading syringe is filled with liquid, and then place the jar (or another heavy object) on the platform and lift it again. Use a ruler to measure how high your load (the jar with water) is moving. (You might need a helper for this step.) Write this distance down.
- You just tested a design in which the helping and leading syringes are equal in size. Switch the helping syringe (the syringe that holds the weights) to one that is larger than the leading syringe. *Do you think lifting the jar will be easier or harder with this design? Do you think the jar will move up more, less, or about the same? Why do you think this?*
- Repeat the procedure with the larger helping syringe to find out! (Notice you will not have to press the helping syringe completely to fill the leading syringe.)
- Use the ruler to measure how high the plate is being moved, and write down that distance.
- Compare the measurements you made for each design. *Why do you think they are different?*
- *Did one design make it easier to move the jar? Why do you think this is so? Which design would you choose to lift a heavy truck—and why?*

EXTRA

Try to lift heavier and heavier objects until you barely can lift them. Remember to place the objects centered on the plate (if a heavy weight is placed off center, the tape might give way and your object might fall). Try this for each configuration and, if you have a scale available, weigh the different weights you were able to lift with each syringe combination. *Which combination was able to lift more weight? Why do you think that is?*

EXTRA

Measure the radius of the small and large helping syringe cylinders. *Can you use these numbers to find the factor by which the weight you can lift gets multiplied when you move from a small to a large syringe?*

EXTRA

Try other combinations of syringe sizes. *Can you see a pattern in your findings?*

EXTRA

Make a hydraulically operated wheelchair lift, dental office chair, or car lift for toys.

OBSERVATIONS AND RESULTS

You probably found it easier to lift the jar with the large helping syringe. In that combination, the jar probably also moved a much shorter distance.

When you press one plunger, the liquid inside the closed system presses with an equal amount of pressure against the other plunger. Because pressure is created by applying force over an area, a small push on a small area creates the same pressure as a much larger push on a much larger area. That is why a hydraulic lift can multiply a force: you push on a small plunger, the pressure that is created moves through the liquid to a much larger plunger, and the load on the larger plunger feels a much greater push than the one you first made. For that reason, it was easier and, if you tried, you could move more weight with a large helping syringe compared to a small one.

Did you see how liquid was transferred from one column to the other in the process? Because the volume of liquid does not change, the larger plunger will not move as much as the smaller plunger. But that is the tradeoff for being able to lift a heavy object!

CLEANUP

Use the water to water some plants, if available. Throw away any garbage. Return all materials to the places where you found them.

Lighten the Load with a Pulley

HEAVE-HO NO MORE: BUILD A SIMPLE MACHINE THAT CAN HELP YOU DO THE HEAVY LIFTING.

Before the industrial age, people largely had to rely on muscle power to move and lift heavy objects. Simple machines such as pulleys, levers, and ramps made it easier for humans to lift heavy objects like rocks and logs. In this project, you will use simple household materials to explore one of these classic machines: the pulley.

PROJECT TIME

15 to 20 minutes

KEY CONCEPTS

Physics
Simple machines
Forces
Energy
Friction

BACKGROUND

Pulleys are simple machines that can make the job of lifting objects easier. They are made from another type of simple machine—the wheel and axle. A simple pulley has one more part added—a rope wrapped around the wheel. One end of the rope can be tied to a load, for example, a bucketful of water at the bottom of a well. You can pull on the other end of the rope and the tension in the rope will lift the load. The advantage of a simple pulley is that it allows you to change the direction of the force required to lift it. For example, to lift the bucket up out of the well, you can pull down on the rope.

What if you want to change the force required to lift a load? With a simple pulley, the force you must exert is equal to the weight of the load. With a compound pulley—two or more pulleys combined together—you can reduce the force required to lift the load. This reduction doesn't come for free though—due to conservation of energy, it increases the distance over which you must exert a force. (Energy equals force times distance.) For example, in order to reduce the force, you must exert to half the load's weight, then in order to lift the load 1 foot (0.3 m), you will need to pull the rope 2 feet (0. 6 m).

There's one more factor you have to consider when using a pulley—friction. An "ideal" pulley—the kind you see in a physics textbook or homework problem—doesn't account for this. Real-world pulleys will always have some friction, and you need to exert a little extra force to overcome it. In this project, you will investigate simple and compound pulleys—and how they change the force required to lift a load.

MATERIALS

- Two empty cereal boxes
- Two pencils
- Four paper clips
- At least 10 metal washers or hex nuts of the same size
- String
- Scissors
- Ruler (optional)
- Empty spool of thread (optional)

PREPARATION

- Stand the two cereal boxes up parallel to each other on a table or desk.

- Carefully poke two holes across from each other toward the top of the cereal boxes (on their inner faces) so you can push a pencil through the holes in a way that will be supported by the boxes. (If necessary, have an adult help with this part.)
- Cut a piece of string that is slightly longer than one cereal box is tall.
- Tie a paper clip to each end of the string.

PROCEDURE

- Hang the string over the pencil with one paper clip dangling off each side. This is your simple pulley.
- Hang two washers from one of the paper clips. (Bend the paper clip slightly so you can easily use it as a hook.) These washers are your "load." Let them fall down to the tabletop.
- One at a time, start hanging washers from the other paper clip. Place the washers onto the paper clip carefully, do not drop them. These washers are your "effort" force. *How many washers do you have to hang on the paper clip until the load starts to lift off the table?*
- Next you will build a compound pulley.
- Poke two more holes (one in each box) so you can insert a second pencil parallel to the first one; they should be a few centimeters apart.
- Cut a new piece of string, roughly twice as long as one box is tall.
- Tie one end of the string to the second pencil, and drape the other end over the first pencil.
- Hang a paper clip from the part of the string between the two pencils so it dangles down between them. Hang two washers from this paper clip and let them fall down to the ground. This will be your load.

- Tie the free end of the string to another paper clip. This paper clip should be up off the ground, close to the pencils. If not, you might need to shorten your string.
- One at a time, start hanging washers from the second paper clip. *How many washers do you have to hang on the paper clip until the load starts to lift off the table?*
- Compare the results for the two different types of pulleys. There should be enough space to set both types of pulleys up at once. If you have enough washers, you can do the tests side by side. *Does it take more or fewer washers to lift the same load using a compound pulley compared with a simple pulley? Why?*

EXTRA

There is a lot of friction in these pulleys because the string rubs directly against the pencil. (There is no true "wheel" that can rotate.) *Can you try to reduce friction in your pulley?* For example, try using an empty spool of thread as a wheel. *How does this change your results?*

EXTRA

Use a ruler to measure the distance traveled by the load and the effort for each type of pulley. *How far must you move the effort force in order to lift the load 4 inches (10 cm) with a simple pulley? What about with a compound one?*

OBSERVATIONS AND RESULTS

You should find that it takes fewer washers to lift an equivalent load using a compound pulley compared with a simple pulley. For example, ideally with a simple pulley you would be able to lift two washers using two other washers. (The effort force would be equal to the load force.) In reality, there is some friction between the string and the pencil that you must overcome, so it might take three or four washers. (The effort force is greater than the load force.) With the compound pulley, ideally you would be able to lift two washers using only one other washer. Again, there is some friction to overcome, so it might take two or three washers. But remember, this decrease in force required is not "free"—you have to move the effort force twice as far. You can confirm this if you measure the distances with a ruler.

CLEANUP

Recycle the empty cereal boxes. Throw away any garbage. Return all other materials to the places where you found them.

Dam Design

WHEN IS IT GOOD IF A DAM SPRINGS A LEAK? WHEN YOU'RE GENERATING POWER! LEARN HOW TO BUILD THE BEST DAM FOR HYDROELECTRIC POWER WITH THIS GUSHER OF AN ACTIVITY!

You may not think much about using water for drinking, cooking, washing, or swimming. But did you know that water can also be used to make electricity? Try this fun activity to learn about hydroelectric power.

PROJECT TIME

20 minutes

KEY CONCEPTS

Potential energy
Kinetic energy
Hydrostatic pressure
Hydroelectricity
Gravity

BACKGROUND

Thanks to gravity, all objects and materials have potential energy when they are raised above the ground. When these objects fall back toward the ground, the potential energy is converted to kinetic, or motion, energy. In the case of falling water, it can be used to turn something called a turbine, which looks like a propeller. If the spinning turbine is connected to an electric generator, it can make electricity. This type of electricity generated by falling water is called hydroelectric power ("hydro" means water).

How do we harness the energy of falling water? To generate a lot of electricity you need a lot of water falling very fast in one place. That means rain is not a good way to generate hydroelectricity, because it is spread out over a very large area and occurs at unpredictable times. Waterfalls have lots of water falling in one place, but because of their natural beauty many are considered natural landmarks and tourist destinations, so people would not be happy if we built ugly power plants on top of them. One solution to this problem is to build dams, or large structures that hold back a stream or river. This causes a deep lake, or a reservoir, to form, and some of that water can be used to generate electricity when it flows through the dam (whereas the rest continues to flow over the falls). In this activity, you will investigate where a turbine should be placed to generate the most electricity.

Note that although hydroelectric power is considered a type of renewable energy, it does have some environmental impacts: the large lakes formed by dams can destroy existing animal habitats; blocking off rivers can mean many organisms such as fish can no longer move from one area to another; and the catastrophic failure of a dam can cause severe flooding and loss of life. If you are interested, you can do your own research on the costs and benefits of hydroelectric power.

MATERIALS

- Milk jug or 2-liter soda bottle
- Small nail
- Tap water
- Duct tape
- Chair, stool, or stepladder
- Bathtub, large sink, or outdoor area where you can spill water
- Tape measure (optional)

PREPARATION

- Gather your materials in an area where you can spill water on the ground, such as outdoors or in a bathtub.

PROCEDURE

- Use the nail to carefully punch three small holes in one side of the milk jug: one near the top, one in the middle, and one near the bottom. Make sure to offset the holes from one another horizontally by a few centimeters. (The holes should not be directly above one another in a straight line.)
- Use pieces of duct tape to tightly seal over the holes.
- Fill the jug with water. Make sure you fill it past the top hole. Look carefully to see if the duct tape seal is holding in the water before you start your test.
- Set the jug on an elevated surface such as a stool or stepladder in an area where you can pour water on the ground.
- Get ready to quickly remove the duct tape. *What do you think will happen when you remove the tape? How will water flow out of the holes? Will the water coming out of the three holes behave the same or differently?*
- Quickly peel off all three pieces of duct tape. Watch closely. *What happens? Where do the streams of water hit the ground?*
- Continue watching as the water drains out of the jug. *Do the streams of water continue to hit the ground in the same place or does it change as the jug drains?*
- Pretend that your milk jug is a dam and the water inside it is the lake it has created. *Where would you place a turbine to generate the most electricity from the water?*

EXTRA

Use a tape measure on the ground to record how far the water gushes out. *How does this distance change if you raise or lower the height of the jug off the ground? How does it vary among the holes?*

SCIENCE FAIR IDEA

Try the activity with different shapes of containers, with holes punched at the same heights from the bottom. *What happens if you compare a milk jug and a 2-liter soda bottle? Does the shape of the container affect how far the water treams out?*

SCIENCE FAIR IDEA

Use different size nails to punch holes (or have an adult help you use a power drill to make holes). *Does the size of the hole affect how far the water flows or how fast the jug drains?*

OBSERVATIONS AND RESULTS

You should observe that the water from the hole at the bottom of the jug gushes out the farthest, followed by the middle hole, then the top one. This is because the hydrostatic pressure (the pressure applied by the water at a certain depth) at the bottom of the jug is the highest, causing the water to rush out at a higher velocity. This high-velocity water has more kinetic energy to turn a turbine and produce electricity, so the bottom location would be the best place to put your turbine.

Depending on where you poked it (and how much you filled the jug), water from the top hole might just trickle down the side of the jug and not actually spurt out horizontally. As the water drains, it will not spew as far from the holes. This happens because hydrostatic pressure is determined by the differences in height between each of the holes' locations and the water's surface—and the surface gets lower as the water drains.

CLEANUP

Recycle any milk jugs or soda bottles used. Throw away any garbage. Return all other materials to the places where you found them.

Candy Snap Materials Science

DID YOU KNOW THAT THERE'S SCIENCE INVOLVED IN SPLITTING A LITTLE TREAT? FIND OUT HOW IN THIS SWEET EXPERIMENT!

Have you ever broken a piece of candy in half to share with a friend? Did you know you were doing a little physics activity when you were snapping that treat? You can learn even more about materials and their properties by playing around with candy in this activity. This is a fun (and tasty) way to study materials science, which includes learning how different materials break—and under what conditions. In this project, you'll divide various candy bars in half and then see what happens when you put them in the freezer. Do you think they'll be easier or harder to break?

PROJECT TIME

75 to 90 minutes

KEY CONCEPTS

Temperature
Physics
Engineering
Materials science

BACKGROUND

Materials scientists and engineers study different materials, how they bend and break, and how to make them stronger. This allows the scientists to help develop stronger, safer structures and objects. Materials can behave differently, depending on whether you stretch, bend, or compress them. If you just stretch or bend a material a tiny bit, it might spring back to its original shape. For example, think about what happens when you bend a metal or plastic ruler slightly. This type of "springy" behavior is called elastic deformation. Almost all materials will show some elastic behavior (even if it's just a tiny bit) before they break.

But what happens if you bend or stretch a material too much? Some materials will continue to deform but permanently hold their bent shape instead of springing back. For example, think about what happens when you bend a paper clip. This is called plastic deformation. Materials that can undergo lots of plastic deformation before they finally break are called ductile. Other materials will break very suddenly when you bend them too much—for example, a wooden ruler or pencil. Materials that snap suddenly without undergoing much plastic deformation are called brittle.

But the behaviors of materials are not always the same. Conditions such as temperature can have a great effect on how they respond. Engineers who build outdoor structures (such as bridges) or things used outside (such as cars) have to worry about how temperature changes will affect a material's behavior. You wouldn't want a structure that's nice and sturdy on a warm day to suddenly collapse if it gets hot or freezing cold! In this project, you'll investigate how temperature affects what happens to different candy bars when you try to bend them in half.

MATERIALS

- Two each of assorted full-size candy bars of your choice
- Two each of assorted "soft" candies such as Twizzlers, AirHeads, or gummy worms (AirHeads are highly recommended because they are the best at showing the experiment's effect.)
- Access to a freezer
- Pen or pencil and paper (optional)

PREPARATION

- Place one of each type of your candy items in the freezer about one hour before you start your experiment.
- Leave the rest of your items out at room temperature. *Do you think temperature will have a large impact on these candy materials' physical behaviors?*

PROCEDURE

- Take one of your room-temperature candy items and hold it on both ends (using both hands).
- Slowly start to bend the full-size candy in half. Pay close attention to how hard it is to bend it. *Is it very stiff and difficult to bend at all? Is it super flexible and easy to bend?*
- Continue bending the candy until it breaks. Pay close attention to how it breaks. *Does it snap very suddenly (brittle) or does it stretch out and break gradually (ductile)?*
- **Note:** Some items, such as Twizzlers and Gummi worms, are very flexible, so you might not be able to break them at all by bending them. *Can you break them by pulling them apart? If so, how does that break occur—suddenly (brittle) or stretchy and gradually (ductile)?*
- If you would like to make a note of each type of candy and its material behavior during bending and breaking, record each observation after you perform the breaking test.
- Now, take the matching candy out of the freezer and repeat the steps above. Remember to carefully observe how the candy breaks. *Is the candy easier or harder to bend? Does the candy break in a different manner?*
- Repeat the procedure for each type of your other candy items. Remember to pay close attention to how hard the materials are to bend and how they break.

- *Did the matching candies behave differently depending on their temperatures? How did different candies of the same temperature compare with one another during the bending test? What do you think would happen if you tried bending these candies when they were warmer than room temperature?*

OBSERVATIONS AND RESULTS

In this activity, you should have seen that most materials became stiffer and less flexible after spending an hour in the freezer. For example, items such as gummy worms and Twizzlers might have been very floppy and bend under their own weight at room temperature but required a little bit of force to bend after being in the freezer. Even if they became stiffer and harder to bend with the cold, they nonetheless might not have experienced brittle breakage. This does not mean your activity went wrong—some materials are still ductile even when they are very cold! Other materials, however, may have shown a very clear transition from ductile to brittle when you froze them. This works very well with AirHeads candies and certain chocolate bars, which are gooey and stretch at room temperature (ductile) but make a very clean "snap" when frozen (brittle).

CLEANUP

Share your leftover candy with some friends!

Slinking Science

Take a Slinky Toy for a Walk

DO ANGLES CHANGE HOW A SLINKY "WALKS" DOWN AN INCLINE? FIND OUT IN THIS FUN EXPERIMENT!

Have you ever watched a Slinky "walk" down a flight of stairs and wondered how it works? It's a fascinating thing to see and a big part of the Slinky's appeal. These spring toys have been popular for more than 75 years; your parents, or even grandparents, may have played with them. Slinkies not only make fun toys; they are also great for doing physics and engineering activities. In this activity, you will investigate how the angle of an inclined plane affects how well a Slinky can walk down that plane. What angle will enable the Slinky to go for the best walk?

PROJECT TIME

30 to 45 minutes

KEY CONCEPTS

Physics
Engineering
Inclined plane
Gravity
Momentum

BACKGROUND

The Slinky is a classic toy in the United States and an interesting tool for physics investigations. It was not originally created, however, with either of these purposes in mind. The inventor of the Slinky was a retired Navy engineer named Richard James, who worked in the Philadelphia Naval Shipyard. He initially thought the bouncy springs would be useful on ships for supporting sensitive instruments and keeping them stable during storms at sea. But after the Navy showed no interest in his springs, his wife, Betty, had a better idea—she thought up a catchy name and Slinky the toy was born in 1945.

So how does the Slinky "walk" down a flight of stairs? To do this, the Slinky slowly flips end over end. If you watch closely, you'll see it stretches to reach the next step down, reforms itself, then stretches again to reach the next step, and so on. This process is possible because of gravity and the Slinky's own momentum. In this way, a Slinky can walk down other surfaces too, such as an inclined plane, which is a type of sloping surface that you will use in this activity.

MATERIALS

- A sheet of plywood. Try to find one that's about 2 feet (0.6 m) long by 1 foot (0.3 m) wide, at least.
- Stack of books
- Protractor
- Stopwatch
- Metal Slinky (full size, not a Slinky, Jr.)
- A piece of paper and a pen or pencil
- A helper
- Other sizes and types of Slinkies (optional)

PREPARATION

- To make your adjustable inclined plane, rest one of the narrow ends of your piece of plywood on a stack of books and the opposite end on the floor.
- Make sure that you have enough books of varying thicknesses to create a 25-degree angle with the inclined plane. This is the largest angle you will be testing. Using a protractor, measure the angle formed where the base of the plane meets the floor.

PROCEDURE

- Adjust the inclined plane so that it is at a 15-degree angle. Use the protractor to confirm the angle formed where the base of the plane meets the floor.
- Have a helper get ready with a stopwatch.
- At the top of the inclined plane hold the Slinky so that it is bent over, making an upside-down U. Let the lower part of the Slinky just barely touch the plane. (This may mean that you are holding around one-half to two-thirds of the Slinky's coils on the upper part of the plane.) *Why do you think you should hold the Slinky this way?*
- Release the Slinky and have your helper time how long it takes the Slinky to reach the bottom of the plane. Also count how many flips the Slinky does during this time. **Tip:** Make sure the Slinky doesn't tumble down the plane but actually flips end over end; do not count any of the "walks" in which the Slinky tumbled.
- *How long did the Slinky take to reach the bottom at this angle, and how many flips did it make?* Write down your results.
- Repeat this at least four more times, each time releasing the Slinky from the same spot at the top of the inclined plane, timing it and counting its flips. *How long did it take in these other trials, and how many flips did it make?* Write down your results.
- Adjust the inclined plane so that it is at a 20-degree angle. Use the protractor to confirm this.
- Repeat the experiment five times at the 20-degree angle, each time releasing the Slinky from the same spot, timing it and counting its flips. *At a 20-degree angle, how long did it take the Slinky to go down the plane, and how many flips did it make?* Write down your results for each of the five walks.

- Adjust the inclined plane so that it is at a 25-degree angle, using the protractor.

- Repeat the experiment five times at the 25-degree angle, each time releasing the Slinky from the same spot, timing it and counting its flips. *At a 25-degree angle, how long did it take the Slinky to go down the plane, and how many flips did it make?* Write down your results for each of the five tries.

- *At which angle did the Slinky make the greatest number of flips on its walk? At which angle did the Slinky take the fastest walk?*

EXTRA

In this activity, you investigated how a Slinky walks down inclined planes at 15-, 20-, or 25-degree angles, but you can change your inclined plane to create a much smaller or greater slope than these and investigate how well the Slinky walks at the more extreme angles. *What are the smallest and largest angles that the inclined plane can be for a Slinky to still be able to "walk" down it?*

EXTRA

There are many ways to use a Slinky to explore different scientific concepts. One activity to try is to put the two ends on the floor and jump. *Can you use the Slinky as a wave detector this way?* Alternatively, try wiggling the Slinky back and forth. *What can you learn about compression waves by doing this?* You can even bang on the Slinky to generate sound. *Can you make different pitches?* Slinkies have also been used as giant antennas. *Can you think of a way to explore using a Slinky as an antenna? What other scientific Slinky uses or investigations can you think of doing?*

SCIENCE FAIR IDEA

A small Slinky, like a Slinky, Jr., might walk down an inclined plane differently than a larger version. Similarly, a plastic Slinky might travel differently than a metal one of the same size. Pick the angle that worked best in this activity and separately try two different types of Slinkies, either different sizes made of the same material or the same size but made out of different materials. *How do the different Slinkies walk down the plane? Does one make more flips or travel faster than the other at the same angle? Why do you think this is?*

OBSERVATIONS AND RESULTS

Did the Slinky do the greatest number of flips on its walk down the inclined plane when it was at a 15-degree angle? Did the Slinky travel fastest down the inclined plane when it was at a 25-degree angle?

Gravity and its own momentum keep the Slinky moving down the inclined plane, and these forces are related to how the Slinky behaves when traveling at different angles. When the Slinky walks down something relatively steep, such as the inclined plane at a 25-degree angle compared with the 15-degree slope, the Slinky should travel faster as gravity pulls it downward. Depending on the exact conditions, this may take about one-third of the time that the Slinky requires at a shallower angle, but the Slinky will not make as many flips on this speedy walk. When the Slinky walks down a surface that is not as steep, such as the inclined plane at 15 degrees compared with 25 degrees, the Slinky should flip more (possibly around two to three times more). When traveling on a 20-degree inclination, the Slinky's speed and number of flips should be in between these two extremes. The exact conditions used in the activity, such as the smoothness of the plywood surface and Slinky's condition, can also affect how the toy performs at the different inclinations.

CLEANUP

Return all materials to the places where you found them.

Duplicate Your Drawings with a Homemade Machine

MAKE A MACHINE THAT WILL COPY YOUR DRAWINGS. AND USE A LITTLE PHYSICS AND ENGINEERING TO MAKE THEM BIGGER OR SMALLER—PRECISELY!

Have you ever wished you could copy a drawing but make it larger or smaller? If you've ever tried to create a larger or smaller copy of your work, you've probably realized that it is hard to get the details right. A machine called a pantograph, could help. It makes copies that can be scaled up or down or made the same size. In this activity, you will make your own pantograph and then duplicate your drawings. Can you figure out how it enlarges, shrinks, or turns drawings upside down?

PROJECT TIME

30 to 45 minutes

KEY CONCEPTS

Physics
Machines
Engineering
Mechanics
Geometry

BACKGROUND

Machines are mechanical devices we use to complete specific tasks. Simple machines, such as levers, allow us to apply force to accomplish a task we might otherwise not be able to complete. (Renaissance scientists defined six types of simple machines: a lever, a wedge, a pulley, a screw, a wheel and axle, and an inclined plane.) Think of a seesaw on the playground. You might not be able to lift your friend off the ground on your own, but if you are each sitting on different sides of a seesaw, you can probably move so that your friend is lifted into the air. This is thanks to the work of a lever.

Machines work by having a power source and a mechanical structure. The power source provides the energy to get the device moving. In the case of simple machines, they were first powered by humans. The mechanical structure controls the movements and makes up most of the body of the machine. In the case of the lever, the mechanical structures include a beam and a pivot point called the fulcrum.

More complex machines combine several simple machines to get a task done. A pantograph is a machine that helps to copy drawings. It combines two levers. How you use it determines if you will get an enlarged, a miniaturized, or a reversed copy. Try it out and see how levers can be used in a surprising way!

MATERIALS

- Cardboard, at least 8 by 3 inches (20 by 7.6 cm)
- Poster board or large cardboard
- Scissors (you might want to have an adult help cut the cardboard if necessary)
- Metric ruler
- Hole puncher
- Marker
- Pencil
- Four brads (round-head paper fasteners)
- Paper
- Tape (optional)
- Adult helper (optional)

PREPARATION

- Carefully cut out (or have an adult help cut) two 8-by-1 inch (20-by-2.5 cm) cardboard rectangles. These will be the arms of the pantograph.

- Carefully cut two 4-by-1 inch (10-by-2.5 cm) cardboard rectangles sized to create the folding part of the pantograph.
- Punch holes near both ends of each of the four rectangles. Punch another hole in the center of each of the larger rectangles.
- Place a longer rectangle in front of you so it forms an "I" shape. Use the other long rectangle and a short rectangle to transform the "I" shape into an "F" shape.
- Use the last short rectangle to connect the end of the short horizontal line in the "F" shape to the middle of the top line in the "F" shape.
- Use a brad and the punched holes to make a joint where the long rectangles overlap.
- Use two brads and the punched holes to make joints in which the short rectangles overlap with the middle of the long rectangles.
- Rotate the short rectangles so the unused holes at their ends overlap. You now have a pantograph that is ready to try out.

PROCEDURE

- Place a poster board on the table. Punch a hole in it near the left border.
- Place the pantograph on the poster board so that one of the unused holes in its arms overlaps with the new hole in the board. Use a brad to attach the end of the pantograph to the board.
- For the first setup, put a pencil through the holes in the overlapping short rectangles near the center of the pantograph. The pantograph now has one unused hole. Place a marker through that hole. Attach the pencil and marker well (use the tape if needed). The tips of the marker and pencil should rest on the poster board.

- Take an unused pen or pencil and draw a 1.5-by-1.5 inch (4-by-4 cm) square on a blank sheet of paper. Place the paper under the pencil and center an empty paper under the marker. In a moment, you will trace the square with your pencil. *What do you think will happen to the marker?*
- Try it out! If the papers slip, tape them to the poster board and try again. *Was your prediction correct?*
- Remove the drawings with squares. Get a new sheet of paper, and draw a 1.5-by-3 inch (3.8-by-7.6 cm) rectangle. Place the rectangle under the pencil and a new blank paper under the marker. (If needed, attach the papers to the poster board with tape.) *What do you think the pantograph will draw when you trace the rectangle with your pencil?*
- Trace the rectangle.
- Remove the new drawing and the one you traced, and compare them. *Is this an exact copy? How is the copy similar and how is it different?*
- Place new papers under the pencil and marker. (If needed, secure the paper to the poster board with tape.) In a moment, you will use the pencil connected to the pantograph to make a small drawing of your choice. *What kind of drawing do you think the marker attached to the pantograph will produce while you are making your own drawing?*
- Try it out. *Was your prediction correct?*
- Try a different configuration of the pantograph. Switch the placement of the pencil and the marker. The pencil you will use to trace or draw is now connected to a long arm of the pantograph, whereas the marker is now placed where the two short rectangles join. *Do you think this change will impact the drawings the pantograph produces?*

- Repeat steps four through ten. You now have three drawings with this setup of the pantograph. *How are these similar and how are they different from your first set of drawings? Can you determine how the placement of the pencil and marker make a difference? Why do you think that is the case?*

- *What do you think makes up the mechanical structure of this machine? What makes it move? What powers it?*

EXTRA

Can you find the two levers in this machine? Can you figure out how they are used in an unconventional way?

EXTRA

You tested two setups of the pantograph. Try a third setup by connecting the middle point of the pantograph (the point that connects the two shorter rectangles of the pantograph) to your poster board with a brad. Put the pencil at the end of one long arm and the marker at the end of the other long arm. *Can you predict what type of drawing will result from this setup?* Try it out!

EXTRA

In geometry, two figures are similar when they have the same shape but not necessarily the same size. The ratio of the length of corresponding sides is called the scale factor. Pantographs are easy ways to create similar figures. *Can you find the factor by which your pantograph scales the drawings for each of the tested setups? How are these two factors related? If you tried the third setup, what scale factor would you say it was?*

EXTRA

Can you make a pantograph that scales your drawing by a specific factor, for example 1/3 or 3, 1/5 or 5, and so on?

EXTRA

Look closely at the details of all your drawings. *Can you find which setup of the pantograph diminishes little defects in the drawing and which one makes them more apparent?*

OBSERVATIONS AND RESULTS

The pantograph should have enlarged your drawings when used in the first setup and made them smaller when used in the second setup. The mechanical structure holding the pencil and marker is made to do just that.

The pantograph you made is a machine. The four pieces of cardboard and the brads are the mechanical structure. It was powered by a human: you! You pushed and pulled the pantograph as you drew or outlined with the pencil. The mechanical structure ensured the marker and pencil moved in unison.

When the end of one arm of the pantograph is fixed in space, the points where the pencil and marker were attached always move in the same direction: when one moves up, the other will move up too, etcetera. The amount they move is scaled because the point farther away from the fixed point always moves more. This allows you to make an enlarged or miniaturized copy.

A third setup, if you tried it, keeps the joint connecting the two smaller rectangles fixed. This forces the two ends of the pantograph (the places where the pencil and marker were attached) to move in opposite directions: when one moves up the other moves down, etcetera. Because both the marker and the pencil were at the same distance from the fixed spot they moved the same amounts, creating an inverted (or upside-down) copy.

If you looked closely, you might have seen how the pantograph uses two levers. Each longer piece of cardboard serves as the rigid beam of a lever that rotates around a joint placed at the end of the beam. As you drew with the pencil, you pushed and pulled on the levers, making the whole pantograph move in an accordionlike motion. Unlike a playground seesaw, however, these levers aren't lifting weight up and down—they are controlling the distance over which an object (in this case the marker) moves.

CLEANUP

Recycle any used cardboard, paper, and posterboard that can't be reused. Return all your materials to the places where you found them.

Balance the Forces within a Mobile

MAKE A MOVING PIECE OF ART—WITH THE HELP OF A LITTLE PHYSICS!

Have you ever seen a mobile, or a hanging art sculpture? You might have had such a mobile in your room when you were little. These mobiles hang in the air and are usually made up of layers of hanging balanced rods, which, in turn, have objects hanging from them. When you look at a mobile sculpture you might wonder how it stays balanced—even when it is in motion. In this activity, you will make your own mobile sculpture and find out.

PROJECT TIME

30 to 45 minutes

KEY CONCEPTS

Physics
Balance
Forces
Gravity
Lever

BACKGROUND

Mobiles are free-hanging sculptures that can move in the air. These sculptures are not only artistic, but they are also a great demonstration of balanced forces. If you look at a traditional mobile more closely, you will usually notice that it is made of various horizontal rods. On each of these rods there are hanging objects attached with a string. When building a mobile, it is important to keep each rod's objects balanced so that the rod stays horizontal and is not pulled down in any one direction. Each rod essentially acts like a seesaw. If you put more weight on one end of it, gravity will pull the rod down in that direction. To balance the rod, you need a counteracting force of the same magnitude on the other side.

It is not only the weight of the objects that matters but also the location on the rod where they are attached. You might notice that not all objects are attached to the rods symmetrically, meaning they are not equally spaced from the pivot point (or the center) of the rod. Even with two objects of the same mass, the rod can be tilted in one direction if one of the objects is farther from or closer to the rod's pivot point than the other one.

A balanced mobile will be at rest if no additional forces act on its structure. When air moves, however, it can push on the objects hanging from the mobile. As a result, they start to move. This motion creates a ripple effect throughout the mobile structure because all its pieces are connected.

You probably now realize that making a mobile not only requires artistic skills but also some knowledge of physics and balanced forces. You will put all of these skills to the test in this activity by making your own mobile. Do you think you will be able to balance all the forces within its structure?

MATERIALS

- Heavy construction paper or cardstock (various colors work well)
- Hole punch
- Pen
- Markers
- Scissors
- Tape
- String
- Straws, at least six
- Ceiling or doorframe you can hang the mobile from (and a chair or adult to help in hanging it)

PREPARATION

- Draw 10 different shapes that you want to attach to your mobile on the construction paper. The shapes should vary in shape and size.
- Carefully cut out the different shapes with your scissors. If you like, you can decorate each of them.
- Punch a hole into the top center of each of the cut-out shapes.
- Attach a piece of string to each of the shapes by threading it through the punched hole and tying a knot. Try to vary the length of string attached to each shape so that they are not all the same.

PROCEDURE

- Start with one layer of your mobile. Attach a piece of string to the center of one of your straws. Hold the straw by the string so it is hanging freely in the air. *Is the straw hanging horizontally? If not, what do you have to do to make it hang horizontally?*
- Once the straw is balanced, tie your first shape to one end of the straw. Again, hold the straw up in the air by its string. *What do you notice happens to the straw?*
- Tie a second shape to the other end of the straw then hold the straw up in the air again. *Is the straw balanced? Why or why not?*
- Balance the straw by moving one of the shapes along the straw. *Can you find a position on the straw where both shapes are balanced?*
- Use a second straw and two more shapes to build another balanced structure. *How easy or difficult is it to balance the two shapes on the straw?*

- Repeat the previous step until you have used up all your cut-out shapes.
- Now put all your pieces together to create a multilayered mobile. Use strings to attach all the straws you made to one another until the mobile is balanced. Use different lengths of string so the shapes and straws do not bump into each other. *Are you able to balance all the forces within your mobile?*
- Once you have successfully balanced your mobile use tape or string to carefully hang it from the ceiling or a doorframe. *When it hangs freely does your mobile move?*
- Carefully blow on one of the shapes hanging from the mobile. *What do you observe?*

EXTRA

You have used many strings with a variety of lengths to build your mobile. Find out if the length of the string that you use to attach your shapes to the straw matters. Make them both the same length and then two different lengths. *Does it change how the straw is balanced?*

EXTRA

Instead of having two different shapes attached to one straw, have one shape on one side and another straw on the other side. Then add more shapes (or straws) to that straw. This is another way to add more layers to your mobile. *How many layers can you build?*

EXTRA

Try testing different lengths of straws. *What changes if you cut one or more straws in half or use straws with different lengths throughout your mobile?*

SCIENCE FAIR IDEA

Can you make a mobile using other materials, such as wooden blocks or other three-dimensional objects? What works best? What are some of the challenges different objects present?

OBSERVATIONS AND RESULTS

Were you able to balance all the forces within your mobile? It might have been trickier than you thought. Even balancing a straw on a string can be challenging. You might have observed that if you didn't attach the string to the very center of the straw, it was pulled down on one side more than on the other. To ensure that the straw hangs perfectly horizontally, the force pulling down on each of its sides must be exactly the same (this will be true if the string is in the center of the straw). When you attached one shape to the straw, the weight of the shape should have pulled the straw down on the side it was attached.

Attaching another shape to the other side of the straw most likely did not fully balance out the straw either. This was because the shape and size of the two attached pieces were probably not exactly the same. The shape that was heavier pulled down on the straw more than the lighter shape. You were probably able to balance the straw out by moving the heavier shape closer to the pivot point (or center) of the straw.

Putting the full mobile together probably took some patience. The mobile probably needed a lot of testing and trying to figure out how to balance the load on each of the straws. You might have noticed that the length of the string didn't affect the balance of the straws very much, as the string itself is not very heavy. When you hung up the mobile and blew on one of the shapes it should have been obvious that the mobile moves with even the slightest breeze. The air movement pushed on the shape and, even though it was a small amount of force, the whole mobile started moving because all of its parts are connected.

CLEANUP

Clean up the remaining construction paper, string, and all your tools. You can hang the mobile up in your room or give it as a gift.

Strength in Numbers Spaghetti Beams

WILL IT BEND OR WILL IT BREAK? LEARN HOW ENGINEERS STUDY MATERIALS FOR USE IN THE CONSTRUCTION OF BRIDGES AND BUILDINGS. BUILD YOUR OWN BRIDGE OUT OF STRANDS OF PASTA AND SEE HOW STRONG IT CAN BE!

PROJECT TIME

20 to 30 minutes

KEY CONCEPTS

Physics
Materials science
Engineering
Tension
Compression

Have you ever helped cook a pot of spaghetti? Strands of spaghetti are pretty long, so sometimes people break them in half so they more easily fit into the pot. How exactly does spaghetti break? And what does this have to do with science? It turns out engineers and materials scientists study how materials break when they are bent. Although professional engineers might be more concerned with steel beams in a bridge, you can do a fun experiment with pasta in your kitchen!

BACKGROUND

What happens when you bend a piece of spaghetti (or another item)? Two things happen when you bend a material: Parts of it are put under tension, meaning they are being pulled apart. Other parts are put under compression, or being squished together. Certain materials tend to break more easily under either tension or compression, so it is important for engineers and material scientists to study how materials break. Better understanding these details allows them to build structures that won't break. Bridges are a great example—when cars drive over a bridge, their weight pushes downward, causing the bridge to bend slightly. This puts the materials in the bridge in both tension and compression. Engineers have to design the bridge to make sure it can handle these forces.

In this project, you will make a "beam" from bundled-together strands of spaghetti. As you hang weights from it, the beam will start to bend—putting some of the strands in tension and some in compression. Which ones break first depends on the physical properties of pasta. Do you think pasta will break first in tension or compression? Try this project to find out!

MATERIALS

- Box of spaghetti
- Two objects of equal height, such as chairs, tables, or large cardboard boxes
- Scissors (and adult's help when using them)
- String
- Paper clip
- Large plastic or paper cup
- Objects to use as weights (such as coins)
- Rubber bands or tape
- Recommended: safety goggles (to protect your eyes from flying fragments of broken spaghetti)

PREPARATION

- Set up two equal-height chairs, tables, or cardboard boxes so they are next to one another, with a gap in between them that is just a few centimeters less than the length of a piece of spaghetti.

- Have an adult help to cut two small holes toward the top of your plastic or paper cup, just under the rim, on opposite sides from one another.
- Tie a loop of string through the two holes in your cup to form a "handle" (turning the cup into a mini-bucket).
- Bend a paper clip into either a "C"- or "S"-hook shape. This will allow you to hang the string handle from strands of spaghetti.

PROCEDURE

- Place a single piece of spaghetti across the gap between your chairs (or tables, boxes, etcetera).
- Hang your cup from the strand of spaghetti using the paper clip hook that you made.
- Slowly start adding weight (such as coins) to the cup. *How much weight do you think the cup will hold before the strand of spaghetti breaks?*
- Continue adding weight slowly. If you are using heavy items such as coins, support the cup with your hand when you drop in a coin, then gently lower the cup until the string pulls on the spaghetti. (Do not just drop or throw coins into the cup unsupported, as this could cause the spaghetti to break more easily.)
- Keep adding weight until the strand of spaghetti breaks. (If it bends and falls through the gap without breaking, move your chairs closer together and try again.)
- Now, bundle together five pieces of spaghetti. Do this by wrapping their ends in either rubber bands or tape to hold them together. *How much weight do you think the bundled strands of spaghetti will hold? Will it be approximately five times the amount a single strand could hold, more than that, or less?*

- Place the bundled strands of spaghetti across the gap and repeat the weight test. Remember to add weight slowly. Watch and listen to what happens as you add weight. *Can you see or hear any individual strands of spaghetti snap before the entire bundle breaks? Are the first strands that break at the top or bottom of the bundle?*

- Try the test again with a bundle of 10 strands of spaghetti (or more, if you have a large cup and lots of coins).

- Observe carefully where the strands that broke are located. *Are the strands that fracture first at the top or bottom of the bundle? Do you think these are are in tension or compression?*

EXTRA

Does how you bundle the spaghetti strands together affect how much weight they can hold? What happens if you loosely tie them with string (just enough to hold them together) versus tightly wrapping them with rubber bands or tape?

EXTRA

Continue the test with larger bundles of spaghetti. You might need to use a larger container (such as a bucket instead of a plastic cup) to break the larger bundles. You can also perform this activity using water as a weight—just be sure the setup is located outside or where spills can easily be cleaned up.

OBSERVATIONS AND RESULTS

You should find that the spaghetti strands toward the bottom of your bundle start to break first. These are the strands that are under tension (being pulled apart). Dry pasta is brittle, meaning it tends to break very rapidly instead of bending permanently (as opposed to a ductile material such as clay, which can be stretched a lot and will change its shape before it breaks). So, when one piece of spaghetti breaks, all the other pieces might follow in quick succession. This type of brittle failure is what engineers want to avoid in structures, such as bridges. Many other factors also go into bridge design to keep them safe and standing—even under heavy weight.

CLEANUP

Throw away any broken spaghetti and other garbage. Return all materials to the places where you found them.

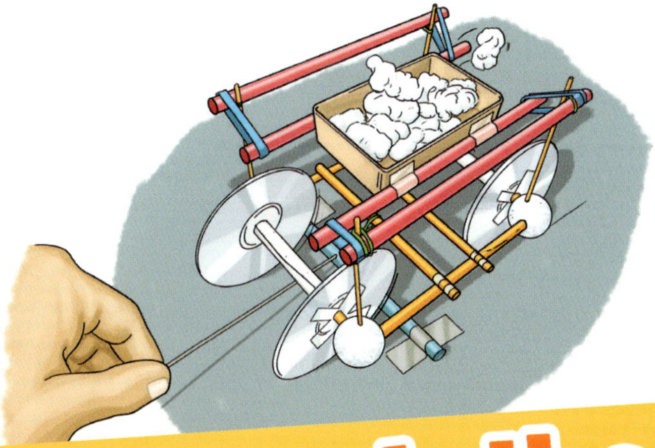

Absorb the Shock!

STOP THE BOUNCE—WITH ENGINEERING! LEARN HOW TO CREATE EFFECTIVE SHOCK ABSORBERS TO KEEP A PRECIOUS CARGO OF PACKING PEANUTS FROM GOING FLYING!

Have you ever ridden in a car over a pothole or a speed bump? You might feel the bumps and get tossed up and down in your seat a little bit, but not nearly as much as you would if the car did not have suspension. Try this engineering project to learn how suspension can help give you a smoother ride!

PROJECT TIME

1 to 2 hours

KEY CONCEPTS

Engineering
Physics
Design
Vibration

BACKGROUND

You might not give it much thought if you are riding down a smooth, nicely paved road, but a car's suspension is very important when driving on a bumpy road or over obstacles. It helps protect the car (and the passengers) from damage when the car goes over a big bump. A car's frame is not rigidly connected to the wheels. If it were, every single bump or vibration experienced by the wheels would be transferred directly to the car and the people inside—making for quite an uncomfortable ride. Instead, the wheels are connected to the rest of the car with a combination of springs (metal parts that bounce back to their original positions when stretched or compressed) and dampers (parts that use friction to slow down motion, usually in the form of a piston filled with viscous oil), also called shock absorbers. Combined, these parts help absorb and decrease road vibrations. In this project, you will build your own simple model car and design a suspension system to help prevent the cargo (packing peanuts) from being tossed out of the car when it drives over bumps!

MATERIALS

- **Note:** This is an engineering design project, which means that there is not an exact list of materials that you must use. The following list is a suggestion, but feel free to substitute other materials.
- Office and craft materials to build a simple car. (If you do not want to build your own car from scratch, you could use a medium-size toy car.) You can search online for "balloon-powered car" or "rubber band car" and you will find many design ideas.

In general you will need:

- Something to use as the car's wheels, such as old CDs or plastic bottle caps
- Something to use as axles, such as wooden skewers or pencils
- Something to use as the car's frame, such as a plastic bottle, cardboard tube, or straws
- Glue or tape to attach the parts together
- Small cardboard box that will fit on your car
- Packing peanuts
- String

- Assorted rubber bands and/or small springs (You can get springs by taking apart ballpoint pens—after asking for permission, of course.)
- Straws or pencils
- Tape
- Scissors

PREPARATION

- Assemble the body of your car. (Skip this step if you are using a toy car.)
- Assemble the axles so they can rotate—for example, thread wooden skewers through straws.
- Attach wheels to the axles. For example, carefully poke holes in the center of plastic bottle caps, then poke wooden skewers through the holes.
- Build a frame for the car and attach the wheels and axles. Make sure the axles can rotate freely and without too much resistance.
- Tie a string to the front of your car so you can easily pull it along.
- If your small cardboard box has flaps, carefully cut them off so its top is open.
- Use tape to attach the box to the top of your car and fill it to the brim with packing peanuts.

PROCEDURE

- Set up "speed bumps" for your car to drive over. For example, tape several straws or pencils to a tabletop, parallel to one another and a few inches apart.

- Pull the car quickly across the speed bumps. *What happens?*
- Now try to design a suspension for your car to prevent the packing peanuts from falling out of the box. For example, you could suspend the cardboard box from rubber bands. This is where the engineering design process really comes into play!
- Once you have built your suspension, try pulling your car over the speed bumps again. Try to do it at the same speed. *What happens this time? Do fewer packing peanuts fall out?*
- Don't get discouraged if your suspension doesn't work on the first try! The engineering design process is iterative, meaning engineers often go back and redesign, rebuild, and retest their devices to improve them. There are many different things you can tweak about your design. For example:

    • Try changing the type of rubber bands. *What happens if you use thicker or thinner rubber bands, or longer or shorter ones?*

    • Try changing the tightness of the rubber bands. *What happens if the box is hanging from them loosely or if the rubber bands are pulled tight?*

    • Try changing the weight of different parts of the car. (Weight has a big impact on vibrations.) *For example, what happens if you tape a bunch of coins to the cardboard box to make it heavier? What happens if you tape the coins to the frame of the car instead?*

- Keep iterating and tweaking your design. *Can you get to a point where no packing peanuts fall out of the box?*

OBSERVATIONS AND RESULTS

You should find that when you attach the cardboard box directly to the frame of your car, many of the packing peanuts are tossed out of the box when you pull it over the speed bumps. This occurs because the vibrations from the wheels are transferred directly to the cardboard box. This would be like riding in a car with no suspension—quite an uncomfortable ride! When you add a suspension, it helps absorb some of the vibrations, so the cardboard box does not bounce up and down as much and doesn't toss out as many of the packing peanuts. It might take some tweaking, however, to get your suspension working well. The stiffness (thickness/length/tightness of the rubber bands or springs) and weight of an object (for example, adding coins) have a big impact on how it reacts to vibrations. Real engineers carefully design a car's suspension to optimally absorb vibrations on the road, just like you did in this project!

CLEANUP

If you built your own car, take it apart and recycle what you can, if you don't plan on keeping the car. Throw away any garbage. Return all materials to the places where you found them.

Surviving the Stormy Seas—with Science

MAKE WAVES: USE A LITTLE PHYSICS AND ENGINEERING TO SEE IF YOU CAN KEEP A HOMEMADE BOAT AFLOAT—EVEN IN ROUGH WATERS!

It's one thing to build a boat that can stay afloat in calm water—but what about one that can survive large waves in a storm? In this activity, you'll put a new spin on a classic activity. You'll build aluminum foil boats to carry pennies as cargo, but instead of testing them in flat water you'll create your own storm to test them in waves!

PROJECT TIME

30 to 45 minutes

KEY CONCEPTS

Physics
Mass
Volume
Density
Engineering

BACKGROUND

People have used boats for transportation for thousands of years. Boats, however, can be vulnerable to storms, when high winds and large waves cause boats to flip over or fill with water and sink. How can you make a boat safer and less likely to sink in water with big waves?

To understand what makes a boat float or sink in still water, you first need to know a little bit about mass, volume, and density (mass per unit volume). An object in water is pulled down by its own weight and pushed up by the buoyant force exerted by the water. If the density of the boat (calculated using the total volume and mass of the boat and everything inside it, including air and cargo) is greater than the density of water, the boat will sink. If the density is less than water, the boat will float. That's why boats can be made of materials (such as metal) that are denser than water: the air space inside the hull helps the boat float. But add too much weight (whether from cargo or water), and it will sink.

Waves can make the situation more complicated. Waves that break over a boat's sides can start filling the boat with water, increasing its density and eventually causing it to sink. So even if a boat could hold a certain amount of cargo in perfectly still water, that doesn't mean it can safely carry that cargo across an ocean filled with waves. And that's not the only challenge. A single, large wave can cause a boat to become unstable and flip over. If you rock a boat a little bit, it should return to its upright position. But rock the boat too much, and it can flip.

Can your boat survive the stormy seas? Try this activity to find out!

MATERIALS

- Aluminum foil
- Pennies or other small, heavy objects to use as weights, such as pebbles, nuts and bolts, and so forth.
- Large container such as a plastic bin or large food storage container. You can also use a sink or bathtub.
- Water (enough to fill your large container)
- Heavy ball (or other object you can drop into the water to create waves next to your boat)
- Location that can tolerate some spilled water (such as bathroom or outdoors)
- Towels for cleanup (optional)

PREPARATION

- Fill your container with several inches of water, and ensure it is in a location where it's okay to spill some water. Bring towels for cleanup if necessary.

PROCEDURE

- Try folding a piece of aluminum foil into a boat shape. There is no single "correct" way to make a boat. (If your first boat doesn't work, you can always make another one.) You will eventually be adding pennies (or other weights) to the boat, so pick a design that will allow your boat to carry this "cargo."
- Place the boat in the water (without any weights), and make sure that it floats. If your boat doesn't float, try modifying it or changing the design. For example, make sure the aluminum foil is folded/crimped tightly. If there are cracks or gaps, water may seep in.
- Try adding pennies to the boat one at a time until it sinks. *How many pennies can the boat hold?*
- Remove the pennies and empty your boat of any water.
- Place the empty boat back in the water.
- Now try dropping your ball into the water from about 1 foot (30 cm) above the water's surface, so that it makes waves. (Make sure you drop the ball to the side of your boat and not directly on top of it.) *How big are the resulting waves? What happens to your boat?*
- Try dropping the ball from increasingly larger heights. *Do the waves get bigger? Does your boat stay afloat?*
- Start over and add a few pennies to your boat, and then try dropping the ball again, starting with the ball back at the lower height. Keep increasing the height of the ball until the boat sinks.

- Keep repeating this process—and adding more pennies each time—until your boat no longer floats at all. *How does the number of pennies affect the boat's ability to withstand waves?*

EXTRA

Try building multiple boats with different shapes. *Do some designs work better than others?*

EXTRA

Do this activity with friends and have a competition to see who can build the sturdiest boat. *What happens if you test all the boats at once—do they crash into one another?* That's another thing that can cause real boats to sink!

SCIENCE FAIR IDEA

Try using other materials to build your boat. For example, make a raft using corks, rubber bands, and wooden craft sticks. *Do the same materials and designs hold the most weight and stay afloat in the biggest waves?*

OBSERVATIONS AND RESULTS

You might have found that there are two different ways your boat can sink in this activity. A single, very large wave can cause a boat to flip over. You probably had to drop the ball from very high to cause this to happen (if it happened at all). Even smaller waves, created by dropping the ball from a lower height, however, can cause a boat to sink. They can go over the sides of the boat, gradually filling it with water (and increasing its density as they replace air in the boat), eventually causing it to become too dense. Just because your boat could hold a certain number of pennies in still water did not mean it could hold the same number of pennies in waves. A boat that is overloaded with pennies—and already riding very low in the water—is much easier to sink like this. As you increased the number of pennies, you probably found that your boat was less able to tolerate waves. The same is true for real boats, which is why they have limits on the amount of people or cargo they can safely carry—whether the seas are calm or choppy!

CLEANUP

Use towels to clean up any spilled water. You can recycle your aluminum foil boats.

THE SCIENTIFIC METHOD

The scientific method helps scientists—and students—gather facts to prove whether an idea is true. Using this method, scientists come up with ideas and then test those ideas by observing facts and drawing conclusions. You can use the scientific method to develop and test your own ideas!

Question: What do you want to learn? What problem needs to be solved? Be as specific as possible.
Research: Learn more about your topic and refine your question.
Hypothesis: Form an educated guess about what you think will answer your question. This allows you to make a prediction you can test.
Experiment: Create a test to learn if your hypothesis is correct. Limit the number of variables, or elements of the experiment that could change.
Analysis: Record your observations about the progress and results of your experiment. Then analyze your data to understand what it means.
Conclusion: Review all your data. Did the results of the experiment match the prediction? If so, your hypothesis was correct. If not, your hypothesis may need to be changed.

GLOSSARY

brittle: Easily broken or cracked.

compression: A force that squeezes something together to make it smaller.

deform: To change something's shape by twisting or pushing it.

density: The amount of mass per unit of volume.

ductile: Having to with being able stretched easily without breaking.

energy: The force that causes an object to move.

force: A push or pull on an object.

friction: The force that resists one object sliding over another.

gravity: The force the pulls objects towards Earth's center.

hydraulic: Having to do with being moved by liquid, like water or oil.

hydroelectric: Power that is created by moving water.

machine: A device the does a physical task.

mass: The amount of matter that makes up an object.

momentum: The measurement of mass in motion.

pivot: To turn or rotate.

tension: The amount something is stretched.

volume: The amount of space an object takes up.

ADDITIONAL RESOURCES

Books

Maslyk, Jacie. *The Fascinating Engineering Book for Kids: 500 Dynamic Facts.* Emeryville, CA: Rockridge Press, 2021.

Oxlade, Chris. *Engineering for Curious Kid*s. London, UK: Arcturus Publishing Ltd., 2022.

Smibert, Angie. *Engineering.* Mankato, MN: The Child's World, 2021.

Websites

Discovery Education
sciencefaircentral.com

Exploratorium
www.exploratorium.edu/explore/activities

Science Buddies
www.sciencebuddies.org/science-fair-projects/project-ideas/list

Science Fun for Everyone!
www.sciencefun.org/?s=science+fair

Videos

Ask an Engineer
https://wqed.pbslearningmedia.org/resource/eng06.sci.engin.design.askengineer/ask-an-engineer/

What Is the Engineering Design Process?
https://wqed.pbslearningmedia.org/resource/phy03.sci.engin.design.desprocess/what-is-the-engineering-design-process/

Publisher's note to educators and parents: Our editors have carefully reviewed these websites to ensure that they are suitable for students. Many websites change frequently, however, and we cannot guarantee that a site's future contents will continue to meet our high standards of quality and educational value. Be advised that students should be closely supervised whenever they access the internet.

INDEX